The Lothians' Last Days of Colliery Steam
Tom Heavyside

The substantial surface buildings at Polkemmet Colliery, Whitburn, sighted from the footplate of Andrew Barclay 0-6-0ST works No. 2358, NCB No. 25, built in 1954, on 23 May 1974. Three other 0-6-0STs from the same maker are also in view.

© Tom Heavyside, 2015
First published in the United Kingdom, 2015,
by Stenlake Publishing Ltd.
www.stenlake.co.uk
ISBN 9781840337150

The publishers regret that they cannot supply copies of any pictures featured in this book.

ACKNOWLEDGEMENTS

I am much indebted to my fellow photographers, mentioned individually in the photographic credits, whose timely images have enhanced this volume. My sincere thanks are also due to Paul Abell, Bob Darvill, Valerie Gillies, Barry Hoper (Transport Treasury), Ellie Swinbank (National Mining Museum Scotland) and the staff of Wigan Heritage Service.

PHOTOGRAPHIC CREDITS

Adrian Booth: page 42 upper.
Tom Heavyside: front cover, inside front cover and pages 1, 4-7, 10-11, 30, 33-38, 40-41, 43-46, 48, inside back cover, back cover.
Douglas Hume: pages 15, 42 lower.
Bill Roberton: pages 31-32.
W.A.C. Smith/Transport Treasury: pages 8, 13, 16, 19-20, 24-25, 28, 47.
Hamish Stevenson: pages 2, 9, 12, 14, 17-18, 21-23, 26-27, 29, 39.

FURTHER READING

Ian Brodie, *Industrial Locomotives of the Lothians*, Stenlake Publishing Ltd, 2006.
Tom Heavyside, *Scotland's Last Days of Colliery Steam*, Stenlake Publishing Ltd, 2003.
Guthrie Hutton, *Mining The Lothians*, Stenlake Publishing Ltd, 1998.
Guthrie Hutton, *Scotland's Black Diamonds*, Stenlake Publishing Ltd, 2001

Regarding the overhaul and maintenance of its locomotives the National Coal Board was virtually self-sufficient - much of this necessary work being undertaken at the various area central workshops. The Newbattle workshops at Newtongrange normally looked after locomotives based in Midlothian, and an impression of the high standard of its workmanship can be gleaned from this photograph taken at Lady Victoria Colliery, Newtongrange, on 16 April 1968. Andrew Barclay 0-6-0ST works No. 1175, NCB No. 8, built in 1909, in virtual pristine condition, had not long been released back into traffic after a visit to the shops.

INTRODUCTION

The counties of East, Mid- and West Lothian effectively encompass the eastern half of the Central Belt of Scotland. Over the years countless millions of tons of coal have been extracted from beneath their surfaces, while from the mid-nineteenth century steam locomotives played a significant role in moving the black diamonds from the pit heads to myriad consumers.

At the dawn of January 1947, under the Coal Industry Act of 1946, thirty-six working collieries spread across the Lothians duly passed into the ownership of the National Coal Board. Included on the inventories prepared by the twenty-one previous colliery owners just prior to Vesting Day were forty-three steam locomotives (usually referred to as 'pugs'), forty carrying their water supplies in a saddle tank, twenty-five with a 0-4-0 wheel arrangement, ten 0-6-0s and five 0-4-2s. The remaining three were 0-6-0 side tanks. As elsewhere in Scotland the products of Andrew Barclay Sons & Co. Ltd predominated, no less than thirty having emanated from their Kilmarnock factory, the oldest back in 1869 while the youngest was less than one year old. A further nine companies were responsible for the other thirteen locomotives. The fleet was dispersed between nineteen locations. In addition two 'Austerity' 0-6-0STs owned by the Ministry of Fuel & Power were based at the Newbattle Opencast Coal Disposal Point.

The importance of the coal industry to the economy of the Lothians can be understood from the facts that in 1947 the NCB employed over 15,800 men across the three counties, and despatched over 4,900,000 tons of coal from the pits. During the 1950s, as the NCB embarked on a much needed modernisation programme, some of the older less productive collieries were closed down, with a number of new drift mines being sunk in order to maintain production in the short term. While by 1960 the total number of outlets had been reduced to twenty-seven, the number of men on the payroll had increased to some 19,000, but with only a marginal increase in the tonnage mined to 5,300,000.

The decade of the 1960s witnessed further pit closures but more notably also saw the start of production at two massive new developments, Bilston Glen at Loanhead, for which the first sod had been cut in May 1952, and Monktonhall at Millerhill, where preparatory work had started in December 1953. By the early 1970s the output from both these collieries was exceeding one million tons per annum, the former employing approximately 2,300 men and Monktonhall some 1,750.

These changes, together with the introduction of more efficient methods of handling the outgoing material, impacted on the steam locomotive stock, such that by March 1967 numbers had been reduced to twenty-seven, eleven 0-4-0STs, eleven 0-6-0STs, four 0-4-2STs and a solitary 0-6-0T. The most elderly was a 0-4-0ST constructed by Grant Ritchie in 1879, while at the opposite end of the age spectrum were two examples newly built for the NCB in 1949 and 1954 respectively, the latter pair the products of Andrew Barclay. In total eighteen of the locomotives carried an Andrew Barclay maker's plate.

After 1967 the run-down of steam continued apace, and following the closure of the landsale yard at Niddrie in the Edinburgh suburbs at the end of 1972, only Kinneil Colliery at Bo'ness and Polkemmet Colliery at Whitburn, both in West Lothian, remained loyal to steam. Perhaps surprisingly steam locomotives continued to be maintained at the latter until 1980 when they were finally usurped by Class 08 diesel shunters hired from British Rail. However it is pleasing to report that no less than seventeen former NCB steam locomotives that ended their days in the Lothians have been preserved, fourteen of which can be found scattered around Scotland, the other three having travelled south across the Border into England.

While the traditional methods of coal mining in Scotland have now been consigned to the pages of history, its importance in the development of the country is commemorated at the National Mining Museum Scotland, based at the former Lady Victoria Colliery, Newtongrange, Midlothian. Further north the much smaller, but equally interesting, Prestongrange Museum, managed by East Lothian Council Museum Services, is centred on the now long-closed colliery of the same name. Steam locomotives are housed at both locations.

Opposite: During the 1950s and early 1960s British Railways owned three important motive depots (sheds as they were commonly known) in Edinburgh: they were well known to railway enthusiasts. Haymarket was the first of the three to close its doors to the iron horse in September 1963 so its resources could concentrate on maintaining the ever-increasing number of diesel locomotives operating in the area. Later Dalry Road shed was abandoned altogether in October 1965, as was St Margarets in April 1967. At the time very few people were aware of the existence of a small shed still servicing steam locomotives in the eastern suburbs of Auld Reekie, just three miles distant from Waverley station. This was the National Coal Board one-road, brick-sided, corrugated iron roofed shed at Niddrie that was not finally closed until the end of 1972. The shed's origins are believed to date from the mid-1870s, the Niddrie Coal Company acquiring its first locomotive in 1875. From 1882 the building was owned by the Niddrie & Benhar Coal Co. Ltd, and in later years, as well as looking after the requirements of Niddrie Colliery, it also serviced the needs of Newcraighall, Woolmet and Edmonstone collieries, although a small shed was available at the former for overnight accommodation. Even after Niddrie Colliery closed in 1927 (Edmonstone had finished the previous year) Niddrie was retained as the company's headquarters where the landsale yard, brickworks and workshops continued in use. Five locos were based at Niddrie when the NCB acquired the site in January 1947, and numbers only declined following cessation of winding at Woolmet in September 1966 and Newcraighall in May 1968. After that, all that remained was the landsale facility that required a meagre amount of work bringing a few wagons from the nearby BR exchange sidings and returning the empties each day. During the last years Andrew Barclay 0-6-0ST works No. 2358, NCB No. 25, new to Niddrie in 1954 with 3ft 6in. diameter wheels and outside cylinders with a 16in. diameter bore and 24in. stroke, was usually entrusted with these far from strenuous duties. Used occasionally was Niddrie stalwart Andrew Barclay 0-6-0ST works No. 1244, NCB No. 19, (formerly 'Niddrie No. 5') that arrived from the makers in 1911 and stayed for the rest of its life. It had similar sized wheels and cylinders as its much younger sister. With its short working day over, No. 25 slowly approaches the shed on the morning of 8 May 1972, the next three photographs also being exposed the same day. Clearly the shed had been extended at some stage, while in the background are some of the other buildings that once belonged to the Niddrie & Benhar Coal Company. Meanwhile No. 19 stood out of sight, silent at the back of the shed.

Right: A front-end view of No. 25 amid heaps of accumulated ash, as it waits to enter the shed.

No. 25 has its saddle tank replenished, the driver standing somewhat precariously on the top rung of the ladder in order to watch the flow of water.

Coaling the locomotives at Niddrie was a rudimentary affair, for with no mechanical plant or other aids available as at many BR sheds, the fuel had to be shovelled by hand from an adjacent wagon. The demarcation between the old and new sections of the shed is clearly evident, with some of the lower brickwork not looking in the best of condition. When steam working ended at Niddrie in December 1972, No. 25 was transferred to Polkemmet Colliery at Whitburn (see page 36), while No. 19 was scrapped on site.

An earlier photograph taken at Niddrie on 20 April 1968. Nearest the camera is Andrew Barclay 0-6-0ST works No. 1233, NCB No. 9, built in 1911, which arrived here in 1961 after overhaul in Newbattle Central Workshops, Newtongrange. Prior to this, apart from a few months at Lady Victoria Colliery, Newtongrange, in the mid-1950s, it had spent almost fifty years at Arniston Colliery, Gorebridge. Behind is Andrew Barclay 0-6-0ST works No. 2026, NCB No. 21, a similar machine to No. 25 illustrated on the previous four pages. No. 21 first touched Niddrie rails as 'Niddrie No. 7' when new in 1937, and except for a relatively short stint at Preston Links Colliery, Cockenzie, in 1963/64, was permanently based at Niddrie throughout its career. Both engines were disposed of for scrap in September 1969. The tall chimney on the left belonged to the Niddrie fireclay works.

The ceremonial cutting of the first sod for the showpiece Monktonhall Colliery at Millerhill was held on 16 December 1953, although it was not until January 1965 that any coal was brought to the surface. Output then increased rapidly so that by the early 1970s the pit was producing over one million tons per annum, most of which was loaded direct into BR high-capacity wagons and transported the six miles to Cockenzie Power Station. What little internal shunting was required at Monktonhall was usually entrusted to a diesel locomotive, but amazingly in 1966 a 0-4-0ST assembled by Shotts Iron Co. Ltd as their No. 6 in 1909, NCB No. 14, was transferred from Newbattle Central Workshops as a spare engine. Records of Shotts locomotive building activity are scanty but the company is known to have put together a few for its own use (probably from parts supplied by Andrew Barclay), five passing to the NCB in 1947, No. 14 then being domiciled at Ramsay Colliery, Loanhead. In the event it appears to have done very little work at Monktonhall where it was photographed in the stockyard on Saturday 13 April 1968, with one of the enclosed winding towers rising above. The shafts reached a depth of 3,051ft. The bold proclamation emblazoned on the accompanying wooden wagon 'Hearts For The Cup' proved wishful thinking, for to the disappointment of many Monktonhall employees, a fortnight later the Edinburgh club Heart of Midlothian was defeated in the Scottish Cup Final 3-1 by Dunfermline Athletic. No. 14 was scrapped in May 1969 as the last surviving Shotts locomotive. Twenty years later the colliery was mothballed by the NCB, only for it to be reopened by a workers' co-operative in 1992. Serious flooding brought an end to the venture in April 1997 and the site was cleared the next year. Monktonhall had the distinction of being the last deep mine in the Lothians.

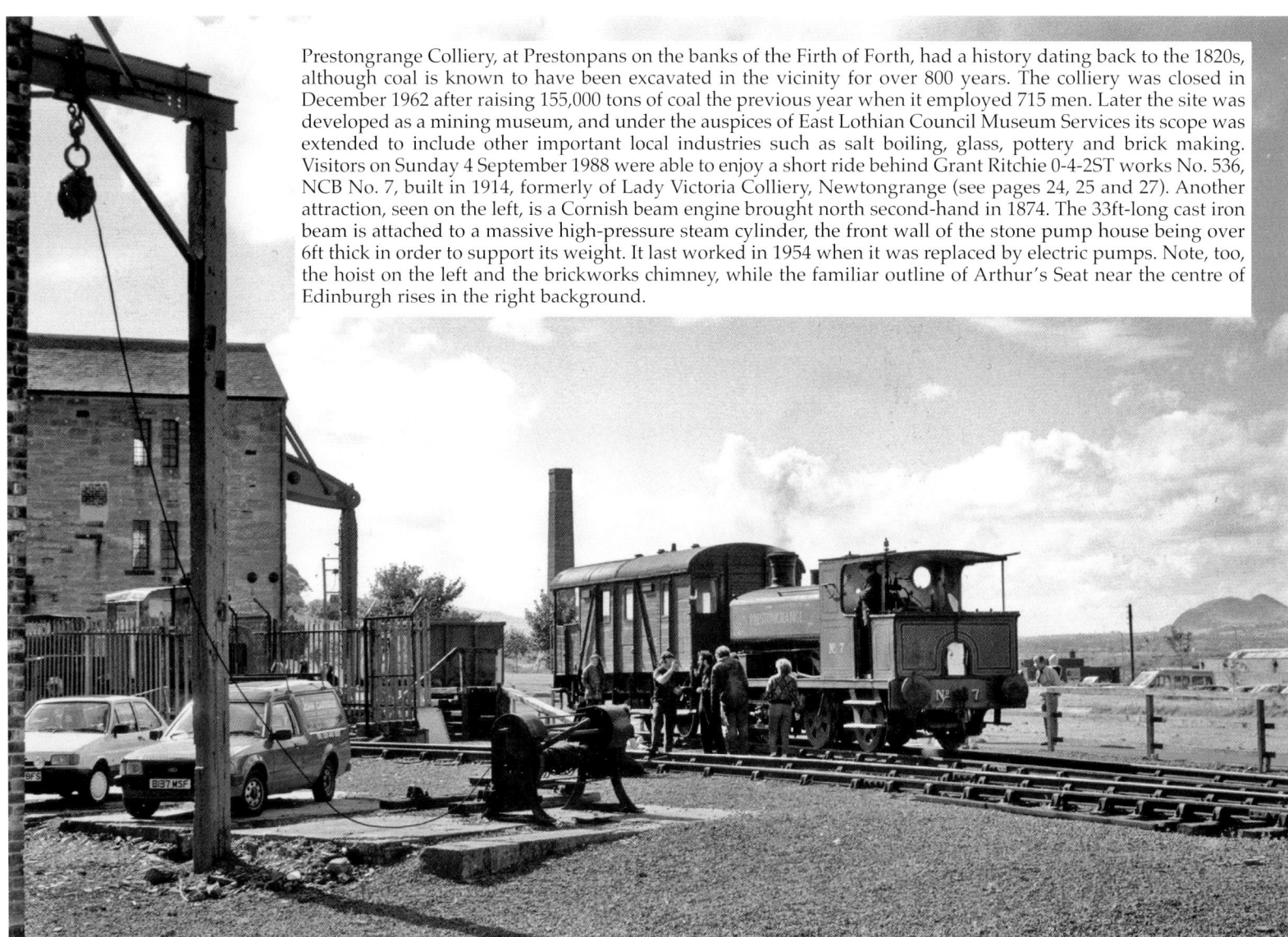

Prestongrange Colliery, at Prestonpans on the banks of the Firth of Forth, had a history dating back to the 1820s, although coal is known to have been excavated in the vicinity for over 800 years. The colliery was closed in December 1962 after raising 155,000 tons of coal the previous year when it employed 715 men. Later the site was developed as a mining museum, and under the auspices of East Lothian Council Museum Services its scope was extended to include other important local industries such as salt boiling, glass, pottery and brick making. Visitors on Sunday 4 September 1988 were able to enjoy a short ride behind Grant Ritchie 0-4-2ST works No. 536, NCB No. 7, built in 1914, formerly of Lady Victoria Colliery, Newtongrange (see pages 24, 25 and 27). Another attraction, seen on the left, is a Cornish beam engine brought north second-hand in 1874. The 33ft-long cast iron beam is attached to a massive high-pressure steam cylinder, the front wall of the stone pump house being over 6ft thick in order to support its weight. It last worked in 1954 when it was replaced by electric pumps. Note, too, the hoist on the left and the brickworks chimney, while the familiar outline of Arthur's Seat near the centre of Edinburgh rises in the right background.

This photograph exposed at Prestongrange on the same day is dominated by a Whitaker Brothers Ltd of Horsforth, Leeds, vertical-boilered crane built around 1890 as their works No. 130. It also came from Lady Victoria Colliery. On the right is Andrew Barclay 0-4-0ST works No. 2219, NCB No. 17, which actually started life here in 1946 after purchase by Summerlee Iron Co. Ltd, the owners of Prestongrange before 1947. It returned here from Newbattle Coal Stocking Site in December 1977 (see pages 14 and 15). The wooden wagon sandwiched between is decorated as belonging to Forth Collieries, which was absorbed by Edinburgh Collieries in 1907. The saddle tank on the left is off the third loco based here, Andrew Barclay 0-4-0ST works No. 1142, NCB No. 29, built in 1908. It spent its last working days on the other side of the Forth at Frances Colliery, Dysart, Fife (see *Fife's Last Days of Colliery Steam*, Stenlake Publishing Ltd, 2014).

The Meadowmill drift mine at Tranent was opened in 1954. It was closed in June 1960 after yielding an average of 62,000 tons of coal per annum, over 200 men being occupied there. The site was then used from 1963 until 1970 as a landsale depot. Standing out of use with its cab boarded up on 4 November 1967 is Andrew Barclay 0-4-0ST works No. 2146, NCB No. 31. Under the grime it will be noted No. 31 is identified as attached to Lothians Northern, for management purposes the Lothians Area being sub-divided into Northern and Southern districts, although as with No. 31 over time this marking became a little meaningless as engines were transferred between divisions. No. 31 was built in 1942 with 14in. x 22in. cylinders and 3ft 5in. wheels and was bought that year by Ormiston Coal Co. Ltd for use at Limeylands Colliery, Ormiston. It was moved to Niddrie loco shed in 1955 and subsequently worked at Fleets (Tranent), Bellyford (Ormiston), and Lady Victoria (Newtongrange) collieries, before gravitating to Meadowmill in 1964. It met its end during the spring of 1970. On the left is the pit-head baths, a comparatively modern-day luxury.

The more powerful Andrew Barclay 0-6-0ST works No. 1833, NCB No. 20, built in 1924 with 16in. x 24in. cylinders and 3ft 6in. wheels makes light work of moving five wagons (a wooden-sided one brings up the rear) during its short stint at Meadowmill on 7 April 1969. The engine had previously spent all its life at Niddrie before its transfer here in September 1968: it moved on to Newbattle Coal Stocking Site two months later in June 1969 (see page 15).

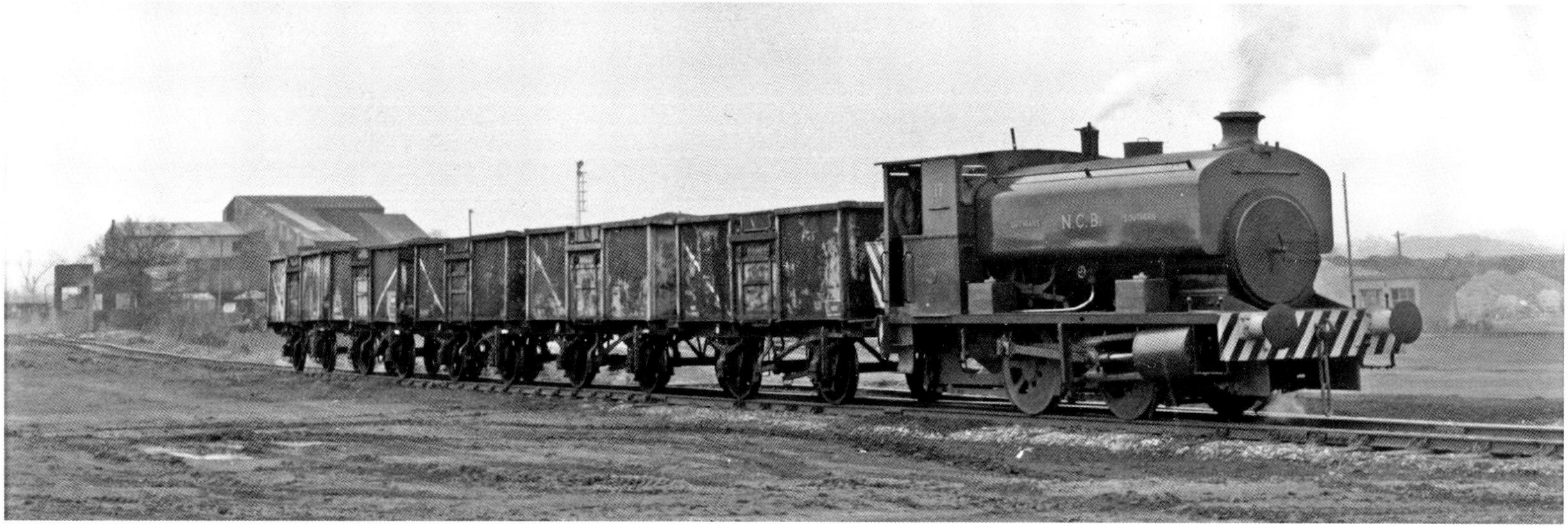

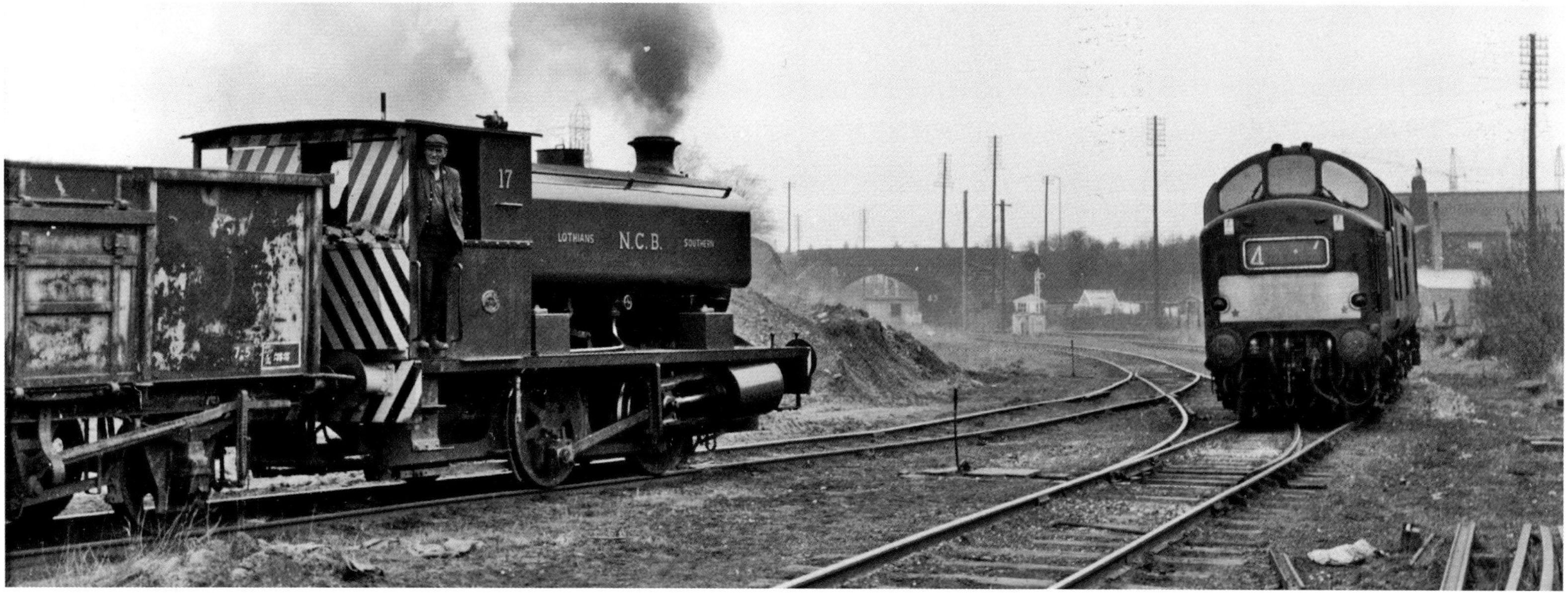

Opposite: In order to take advantage of seams lying comparatively close to the surface, under the management of the Ministry of Fuel & Power opencast mining was commenced at Newtongrange during the later stages of the Second World War. Motive power at the Newbattle disposal point was in the hands of Hunslet-designed 'Austerity' 0-6-0STs, which continued when the operation passed to the newly formed NCB Opencast Executive in April 1952. Excavation of coal ended early in 1961 when the remaining 'Austerities' were dispersed elsewhere. From 1963 the site was taken over by the NCB Lothians Area for coal stocking purposes, known officially as Newbattle but often referred to as 'Butlerfield'. When it opened one of the first locomotives drafted in was Andrew Barclay 0-4-0ST works No. 2219, NCB No. 17, built in 1946, which arrived from the nearby Lady Victoria Colliery. It is seen (above) in quite splendid external condition on a dull 16 April 1968 as it draws five wagons away from the screens while (below), a few moments later, it greets BR English Electric Type 3 (later Class 37) diesel-electric No. D6844 based at Haymarket shed, Edinburgh, arriving to collect some outgoing coal. Note the saddle tank of No. 17 stencilled 'Lothians Southern', the NCB identification number positioned on the side of the cab and the yellow and black diagonal warning stripes across the buffer beams, back cab sheets and bunker. The more elderly of the pair is still extant, at the Prestongrange Museum in East Lothian (see page 11), while the diesel locomotive, released from the maker's factory at Newton-le-Willows, Lancashire, in June 1963, was withdrawn from main line service in January 2000.

Below: Andrew Barclay 0-6-0ST works No. 1833, NCB No. 20, built in 1924, hauls a short rake of loaded wagons at Newbattle on 21 July 1969. It had been moved here from Meadowmill landsale depot (see page 13) the previous month.

Opposite: No. 20 was in exemplary external condition when observed taking water at Newbattle the next month on 25 August 1969. Note it is simply marked as belonging to the NCB Lothians Area. After rail traffic ceased in 1971 No. 20 languished here until autumn 1979, when it was rescued for preservation. Initially it was based in the Scottish highlands on the Strathspey Railway at Aviemore, but in March 2012 it was transported south to the Ribble Steam Railway at Preston, Lancashire.

Right: Sadly awaiting their demise at Newbattle, where they had resided since 1963, are two Kilmarnock-built veteran 0-4-0STs on 28 December 1967. Octogenarian Grant Ritchie works No. 39, NCB No. 2, on the left, was outshopped from their Townholm Works back in 1879, at first being employed in its native Ayrshire before it was bought by the Lothian Coal Company in about 1896. Over the next fifty years it was used at Polton (Bonnyrigg) and Whitehill (Rosewell) collieries. Then under the NCB it saw further service at Whitehill and Arniston Colliery, Gorebridge, from 1959, before being relocated to Newbattle. Its companion, Andrew Barclay works No. 851, NCB No. 28, left the manufacturer's Caledonia Works in 1899 for James Waldie & Sons Tranent Colliery in East Lothian, before the business was absorbed by Edinburgh Collieries Ltd in 1907. While owned by the NCB it worked successively at Wallyford Wagon Works, Meadowmill, Prestongrange and Lady Victoria collieries, prior to Newbattle. No. 28 sports yet another variation of ownership detail, 'Lothians North'. The duo stood forlorn and neglected for a further eight months, until in August 1968 they were both broken down into small manageable pieces and carted away as scrap metal by Motherwell Machinery & Scrap Co. Ltd.

This Andrew Barclay 0-4-0ST, alongside the corrugated iron shed at Newbattle on the same day as the photograph overleaf, is of unknown vintage, its original identity having been lost when rebuilt by the Lothian Coal Company in their Newbattle workshops in 1931. As a second NCB No. 2 from 1947, it spent time shunting at Whitehill and Lady Victoria collieries and Meadowmill landsale depot before its reallocation here about October 1967. In July 1968 it was returned to Newbattle workshops but in September 1969 suffered a similar fate as the discarded examples on the previous page, again at the hands of men employed by Motherwell Machinery & Scrap Co. Ltd.

Opposite: The Lothian Coal Company owned a number of collieries in the Newtongrange area; collectively often referred to as 'Newbattle Collieries'. Work on what became the company's main asset Lady Victoria Colliery, adjacent to Lingerwood Colliery, started in August 1890 and took five years to complete. It was named after the wife of the Marquis of Lothian. During 1947 the NCB employed 940 men at the mine, 820 labouring underground and another 120 on the surface to produce 291,000 tons of coal. During the years 1950 to 1956 the colliery was reconstructed, including laying a 3ft 6in. gauge railway operated by Hunslet-built diesel locomotives, whereby coal raised at Lingerwood and Easthouses collieries could be transported to Lady Victoria for processing. The records for 1960 provide a snapshot of the scale of the operation at Lady Victoria following the modernisation work, Lady Victoria itself lifting 515,000 tons of coal from below ground with 1,793 men on the payroll, while Lingerwood contributed 206,000 tons (718 employees) and Easthouses 466,000 tons (1,088 employees). Here Andrew Barclay 0-4-0ST, NCB No. 2, seen seven months earlier than on the occasion at Newbattle, propels loaded wagons away from the colliery on 18 May 1967. Note the dint in the left-hand cylinder casing from an earlier minor altercation, while a haze hangs over the area in the wake of its exertions. The large two-road engine shed can be seen above the loco.

A panoramic view of the sidings at the south end of 'Lady Vic', as it was often irreverently known, as Andrew Barclay 0-6-0ST works No. 1175, NCB No. 8, built in 1909, equipped with 15in. x 20in. cylinders and 3ft 6in. wheels, leads a set of empties towards the colliery yard on 11 June 1968. Previous to its move here about 1953 Arniston Colliery at Gorebridge had been the engine's sole place of employment. The busy landsale yard can be seen on the left where a lorry waits by the hopper facility, while nearer the camera a couple of men go about the arduous task of filling sacks by hand prior to loading them onto the adjacent lorry. Meanwhile a third lorry sets off on its delivery round. On the right are the tracks of the Edinburgh to Hawick and Carlisle 'Waverley' route, from which passenger services were controversially withdrawn by BR the following January.

During its latter days Lady Victoria became well-known for its use of locomotives with a far from common 0-4-2 wheel arrangement, as fitted to lifelong 'Lady Vic' resident Andrew Barclay saddle tank works No. 1193, NCB No. 6, built in 1910. The detail of the motion on the right-hand side connecting the 3ft 5in. driving wheels to the 15in. x 22in. cylinders, and the smaller trailing wheel, can be digested from this photograph taken on 28 December 1967. The five-plank wagon No. L3 behind the cab was designated for loco coal only. A number of other internal user wagons can be observed, as can the roof of the workshops.

Opposite: Later on the same day as the previous photograph No. 6 was seen contemplating its next move near the weighbridge. The interlaced trackwork over the weighbridge is evident, while a second locomotive can be glimpsed on the far left of the frame shunting wagons over a second set of scales. The notice on the approach specified a maximum weighing speed of 2 miles per hour. Miners were a somewhat hardy lot for as can be seen, even on this very cold but bright December day, fresh air percolates through a couple of open windows of the smoke-blackened overhead cabin, where the tonnage of each wagon passing underneath was meticulously recorded. Also from this lofty position the yard foreman could keep a close watch on most of the goings-on around the site. Pleasingly No. 6 has since become a centenarian, after finding a safe refuge in the late 1970s on the Tanfield Railway at Marley Hill, County Durham.

Below: The distinctive weighbridge arrangement at Lady Victoria is seen from the opposite side as Andrew Barclay 0-6-0ST works No. 1458, NCB No. 3, propels a motley array of wooden-sided wagons of varying sizes around the structure on 26 March 1968. The engine was originally purchased as a wartime need by the Ministry of Munitions for the Houston factory, near Paisley, in 1916, but after becoming redundant there in 1919 following the end of hostilities it was bought by the Lothian Coal Company for Lady Victoria. It enjoyed a short sojourn at Preston Links Colliery, Cockenzie, in 1959/60, but otherwise remained loyal to 'Lady Vic' until taken out of service in 1970. Fortunately No. 3 escaped the breaker's hammer, and these days is on the stock list of the Bo'ness & Kinneil Railway where, in honour of its past, it has been named 'Lady Victoria'.

The Kilmarnock-based company Grant Ritchie, in business from 1876 until 1920, manufactured a wide range of machinery, particularly for the mining industry, including over forty steam locomotives. This 0-4-2ST was delivered to Lady Victoria as works No. 536 in 1914 and stayed there throughout its working life, except for an interlude of about three years from 1954 one mile down the road at Arniston Colliery. Here, against a backdrop of the main colliery buildings as NCB No. 7, the engine awaits its next assignment on 25 August 1969.

Here No. 7 shows its prowess while handling a long line of wagons at Lady Victoria on 16 February 1971.

Nos. 3 and 6 (left) busy shunting at Lady Victoria on a rather overcast 16 April 1968. The tall chimney provides the escape route for the smoke produced by a battery of Lancashire boilers below.

With work over for the day on 26 March 1968, a begrimed No. 7 slowly approaches the shed at Lady Victoria. The BR-manned Lady Victoria Pit signal box on the left, as well as controlling movements along the main line, authorised those to and from the colliery network. The horizontal position of the semaphore signals would indicate a lull in proceedings. Nowadays No. 7 can be viewed at the Prestongrange Museum (see page 10).

Left: The two locomotives depicted on this page at Lady Victoria had already run their last when the photographer called on 25 August 1969. Grant Ritchie 0-4-2ST works No. 527, NCB No. 5, was built in 1908 for the Lothian Coal Company, and except for visits to the adjacent workshops, where it was rebuilt in 1933, never left the site. The motion had already been dismantled as spares for possible use on other locomotives, the rest of the engine being scrapped a few days later.

Right: This 0-4-0ST was constructed in the Lothian Coal Company's Newbattle workshops in 1927 (most of the major components probably being obtained from Andrew Barclay) as their No. 1, a number it retained during its NCB days. It too had had its motion removed, although it was not until the following April that it faced the oxyacetylene torches.

The Lothian Coal Company established some well-equipped workshops at Newbattle, essentially part of the Lady Victoria Colliery complex. It could handle most of the company's day to day maintenance requirements, including any necessary locomotive and wagon repairs. In addition to the 0-4-0ST pictured opposite attributed to the factory, three others are known to have been rebuilt in the shops during the early 1930s. When the NCB was formed the Lothians Area designated the premises as its central workshops, and over the next twenty-five years nineteen steam locomotives and three diesels are recorded as having visited for repairs and overhauls, some on more than one occasion. Standing in the compound outside the workshops, amidst a mass of other items, are a couple of Andrew Barclay 0-6-0STs, works No. 1175, NCB No. 8, built in 1909 (left) and the slightly more senior works No. 1023, NCB No. 23, of 1904, on 28 December 1967. After attention No. 8 was able to resume its duties the next year (see page 2), but No. 23 was less fortunate, being deemed uneconomic to repair and subsequently scrapped by Motherwell Machinery & Scrap Co. Ltd in September 1969.

NATIONAL MINING MUSEUM SCOTLAND

The last rail-borne coal left Lady Victoria Colliery in April 1972, although production was maintained until 1981, the workings below ground having been married up with those of Bilston Glen and Monktonhall. Even before the NCB vacated the site it had been recognised that many of the surface buildings were some of the finest examples of late Victorian engineering practice still in existence and, if possible, needed to be preserved for the benefit of future generations. With the added advantage of being located just seven miles from the centre of Scotland's capital city and close by the A7 trunk road leading south, plans were soon being discussed for the metamorphosis of the colliery as a national mining museum.

Among the original structures that swayed many minds that Lady Victoria would be ideally suited to a new role as a colliery museum was its most identifiable feature, the 85ft-high headstock and accompanying winding wheels (whorls) above the shaft. Of added significance was the fact that Sir William Arrol & Company of Glasgow built the former, the same company who only a few years earlier had been responsible for the gigantic Forth Bridge. Also still in situ was the attendant steam powered winding engine capable of lifting up to 11 tons, supplied by Grant Ritchie & Company of Kilmarnock. Grant Ritchie, as illustrated on previous pages, also built a number of steam locomotives.

Inevitably some of the buildings are from a later age, such as the boiler house which replaced the original in 1924 when twelve Lancashire boilers by Tinkler Shenton & Company of Manchester were installed, after purchase second-hand from a munitions factory at Gretna. These provided the necessary steam for the winding engine up until the end of the colliery's commercial life. A new powerhouse was also erected in 1924 for supplying electricity, like its predecessor, to the three Newbattle pits.

The sanitised Lady Victoria Colliery first opened its gates to the public in 1984, and as the National Mining Museum Scotland it is now one of Scotland's Five Star visitor attractions. Among myriad exhibits of a bygone age, the important role railways played has its deserved place, including Andrew Barclay 0-4-0ST works No. 2284, built in 1949, which is representative of scores of their products that once could be seen at work throughout the Scottish coalfields shifting the black stuff. The engine can be inspected along with many other large items, including a 40ft diameter coal shearer, underground cages and carriages etc., on the guided 'Big Stuff Massive Machinery Tour'.

From September 2015 the museum has become even more accessible with the reopening of a section of the former 'Waverley' route from Edinburgh as far as Tweedbank, the line by which coal once departed from the site (see pages 20 and 27). Newtongrange station is within easy walking distance.

The words of a verse from a poem by Valerie Gillies are most apt:

The Lady is the last of her kind.
Headframe in the clouds, these pulley-whorls
Change with the light, a beacon to remind
Who fuelled Scotland, lit us, kept us warm.

The cast oval identification plate attached to Andrew Barclay works No. 2284.

Arniston Colliery at Gorebridge wound its last coal in April 1962, the washery being retained for a further year before the site was adapted by the NCB as a withdrawn machinery store. Redundant items, large and small, were transported from closed collieries and stored for possible use elsewhere or sold for scrap. Four standard gauge steam locomotives are known to have met their end here in the mid-1960s, and later a number of diesels, including some narrow gauge examples, suffered a similar fate. Andrew Barclay 0-4-0ST works No. 2284, NCB No. 21, built in 1949, pictured amidst an assortment of discarded items in August 1978, arrived by low-loader from Cairnhill drift mine at Cronberry, Ayrshire, the previous December. Photographs of the engine at Cairnhill, and on the Waterside Railway, near Dalmellington, where it was domiciled until autumn 1973, can be seen in *Ayrshire's Last Days of Colliery Steam* (Stenlake Publishing Ltd, 2013). As mentioned opposite, the engine is now part of the National Mining Museum Scotland collection.

Opposite: Prior to the mining interests of the Shotts Iron Company being taken over by the state in 1947, the company had formulated some ambitious plans to deepen Burghlee Colliery at Loanhead in order to exploit reserves estimated to exceed 100 million tons. In the event the NCB decided to sink new shafts half-a-mile to the west of Burghlee at Bilston Glen, where preparatory work started in May 1952, production commencing in February 1961. The last coal was brought up the Burghlee shafts in October 1964. While diesel locomotives normally handled the shunting requirements at Bilston Glen, they were joined in 1965 by Andrew Barclay 0-4-0ST works No. 1996, NCB No. 15, later No. 29, as a spare engine in case of need. The locomotive had originally been purchased new by Edinburgh Collieries Ltd in 1934 and was employed at Fleets Colliery, Tranent, until 1955, subsequently working from Niddrie shed, followed by spells at Limeylands, Meadowmill, Arniston and Lady Victoria collieries before its move to Bilston Glen. No. 29 was seen at the latter dumped in the headshunt on a wet day in September 1977. The main colliery buildings can be seen in the background, the winding gear for the upcast shaft on the far left of the frame and that for the downcast above a set of empty BR wagons. No. 29 lay here undisturbed for another twelve years before becoming a static exhibit in the more genteel surrounds of Pittencrieff Park, Dunfermline, at about the time Bilston Glen closed down in 1989. It was a sad end for a colliery that had once produced over one million tons of coal per annum and employed some 2,300 men.

Below: A panoramic view of Polkemmet Colliery at Whitburn, West Lothian, as septuagenarian Andrew Barclay 0-6-0ST works No. 885, NCB No. 8, built in 1900, climbs the steep gradient to the BR exchange sidings with a couple of wagons in tow on 10 May 1972. Note the vast mound of bing (colliery waste) dumped on the right-hand side of the picture.

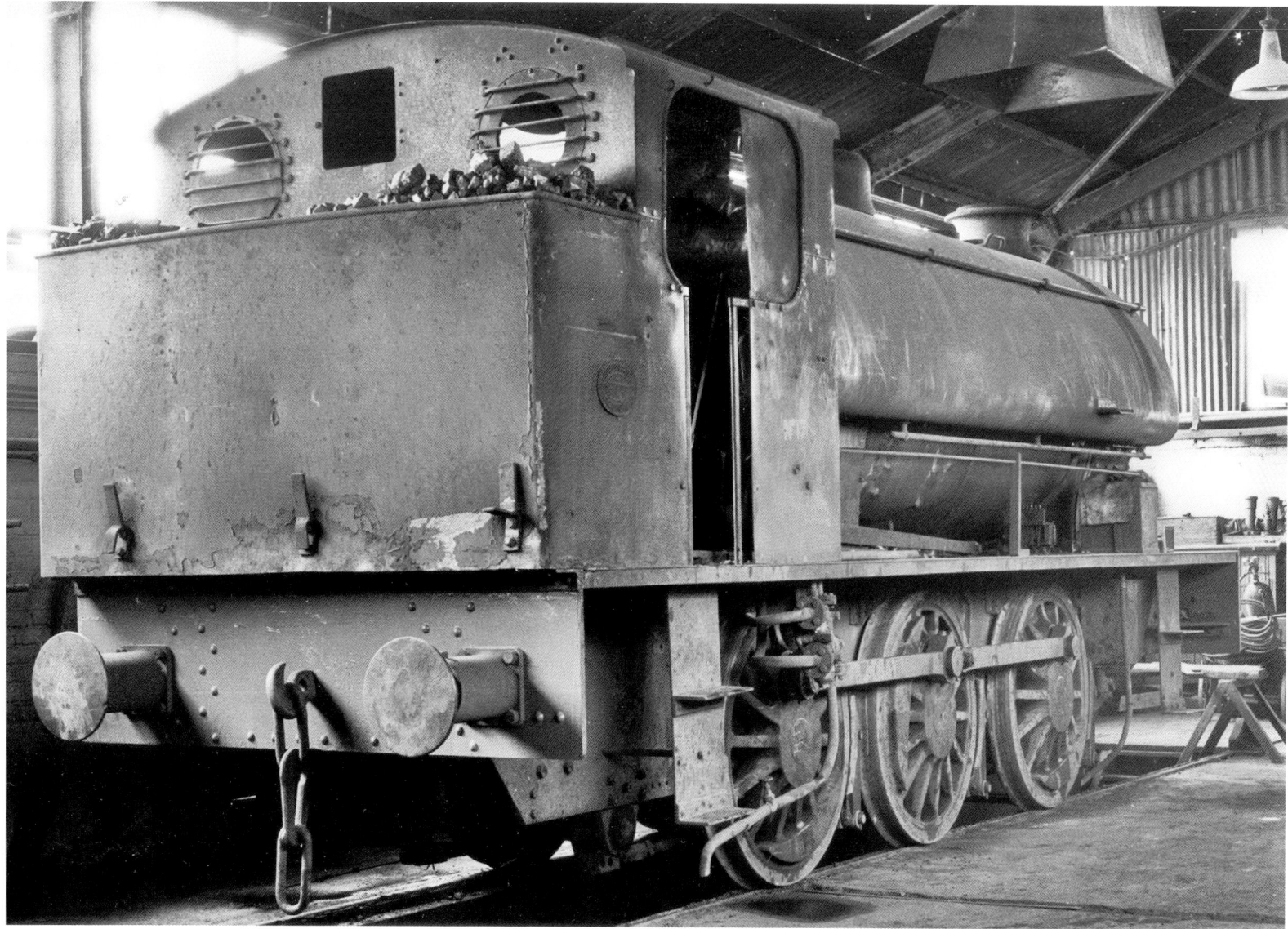

Opposite: William Dixon Ltd, one of Scotland's foremost coal and iron companies, started sinking Polkemmet Colliery just before the start of hostilities in 1914, but with construction work having to be suspended because of the war it was 1923 before it became operational. Dixon's invested in the site to gain access to some high-grade coking coal in the area to supply its own ironworks at Calder, near Airdrie, and at Govan, Glasgow. Later, during the mine's first year under the banner of the NCB in 1947 it raised some 300,000 tons of coal, with 940 men then relying on the pit for their take-home pay, the output increasing to 587,000 tons in 1960 with a corresponding increase in manpower to 1,902. Production levels were still over 500,000 tons per annum in the mid-1970s, although due to increased efficiency the number of employees had been reduced to 1,440. At rest inside the two-road shed on 23 May 1974 is Hunslet 'Austerity' 0-6-0ST works No. 2880, NCB No. 17. The engine was originally ordered by the Ministry of Supply and delivered from the maker's Jack Lane factory in Leeds to the Army depot at Long Marston, Warwickshire, in November 1943. As WD No. 75031 it was moved north to the Central Ammunition depot at Longtown, near Carlisle, in 1956, before being purchased by E.G. Steele Ltd of Hamilton in 1959, who subsequently resold it to the NCB for use at Polkemmet in 1961. This is the only 'Austerity' of the 485 constructed over a twenty-one year period from 1943 to be employed at a Lothians colliery.

Right: Two years earlier the austere yet functional lines of the 'Austerity' design are readily apparent, as No. 17 in filthy external condition hauls some loaded wagons towards the BR exchange sidings on 10 May 1972. This robust machine equipped with 18in. x 26in. inside cylinders powering 4ft 3in. wheels, providing a nominal tractive effort of 23,870lb. at 85% of maximum boiler pressure of 170lb. per square inch, proved highly suited to the heavy nature of the work at Polkemmet. Note the empty wagons behind No. 17 along with another large accumulation of bing in the background. Today in retirement, No. 17 can still be seen in West Lothian as a valued asset of the Bo'ness & Kinneil Railway.

Anyone seeking the sight and sound of steam locomotives hard at work were seldom disappointed by a visit to Polkemmet. Here two Andrew Barclay 0-6-0STs, both comparatively recent additions to the Polkemmet allocation, strain every joint as they lift a dozen wagons up the one mile 1-in-34 gradient from the colliery yard to the exchange sidings on 23 May 1974. The leading engine, works No. 1175, a second NCB No. 8, built in 1909, came from Lady Victoria Colliery in March 1973 (see pages 2 and 20), while its much younger companion, works No. 2358, NCB No. 25, built in 1954, had been transported from Niddrie the previous December (see pages 4 to 7). The wagons congregated on the left are empty, while the loaded wagons in the background await the arrival of a BR diesel locomotive, which will approach the sidings via the branch from Benhar Junction, between Fauldhouse and Shotts on the ex-Caledonian Railway Edinburgh to Glasgow Central route.

Right: A few moments later yet more voluminous black smoke is ejected into the atmosphere. The bleak nature of the surrounds on Polkemmet Moor is evident.

On the same day as the pictures overleaf NCB No. 8 (Andrew Barclay works No. 1175) runs slowly towards the back of the substantial brick-built weighbridge hut. Inside the building the number and weight of each empty wagon was carefully recorded as it passed over the weighing mechanism on the other side of the hut - the scales can be glimpsed through the window. With their details having been documented the wagons in view now await their turn to descend, under the force of gravity, towards the screens in the background for refilling.

This 0-6-0T built by the well-known Leeds-based manufacturers Hudswell Clarke & Co. Ltd had 16in. x 24in. cylinders and 3ft 9in. wheels and was similar to a number supplied by the company to the Port of London Authority from 1915. It left the West Riding as works No. 1331 in 1918 after purchase by William Beardmore & Co. Ltd for its ironworks at Mossend, Glasgow, and moved to Polkemmet in 1935 after sale to William Dixon Ltd. It was nearing the end of its days when photographed at Polkemmet on 30 March 1968 as NCB No. 16, a number retained from its Dixon days, which obviated the need to remove the cast plates from the side of the tanks. Surprisingly it was transported to the modern Killoch Colliery (sunk during the 1950s) at Ochiltree, Ayrshire, in November 1969, only for it to be scrapped there the following April. It had the distinction of being the only engine built by Hudswell Clarke to ever grace the Lothians' coalfields.

Two NCB No. 8s outside the shed at Polkemmet on 23 May 1974, this confusing situation coming about when Andrew Barclay works No. 1175 (nearest the camera) joined Andrew Barclay works No. 885 here in March 1973. The former was allocated running number 8 in the early days of the NCB by the Lothians Area when based at Arniston Colliery, Gorebridge, whereas its stable-mate was renumbered 8 from its old Dixon's No. 14 when part of Central East Area stock sometime before 1967. The rationalisation of the area structure and transfers of locomotives between collieries led to similar situations elsewhere in the 1970s, but by this time often the NCB didn't bother to renumber one of the engines. Some significant differences can be discerned in the design of these two 0-6-0STs built nine years apart, the older more squat example on the right from 1900 being the more powerful. This had 18in. x 24in. cylinders and 3ft 8in. wheels compared to the 15in. x 20in. cylinders and 3ft 6in. wheels of its sister. In December 1977 Andrew Barclay works No. 885 migrated south to the Cambrian Railways Society headquarters at Oswestry, Shropshire.

Above: On the same day as the picture opposite Andrew Barclay works No. 1175 is about to have its saddle tank replenished. In the middle distance, behind the lighting pylon, is the weighbridge where the total tonnage of the loaded wagons was recorded as they gravitated away from the screens towards the despatch sidings. By subtracting the weight of the wagon when empty (obtained earlier) an accurate tally of the coal in each individual wagon could be ascertained and invoiced accordingly.

Below: Later the same day, as the shift draws to a close, No. 25 has its smokebox cleared of accumulated ash.

Right: Yet another No. 8 at Polkemmet! Andrew Barclay 0-6-0T works No. 1296, built in 1915, was renumbered 8 following its move to the Fife Area in 1962, previously running as No. 2 in the Alloa Area. It had 18in. x 24in. cylinders and 3ft 7in. wheels, the oblong Giesl ejector being fitted in 1965 during a visit to the NCB Dysart Central Workshops. It arrived at Polkemmet in April 1977 sporting a dark blue livery with the cab and bunker painted a garish yellow - the house colours of the NCB, having earlier led a somewhat nomadic existence (see *Fife's Last Days of Colliery Steam*, Stenlake Publishing Ltd, 2014). Here it is piloted by fellow No. 8 (Andrew Barclay works No. 1175) on 19 May 1977. Some wag has chalked 'The Flockton Flyer' on the smokebox door of the leading engine, a reference to the popular children's television series first shown in the spring of 1977, the outdoor scenes being filmed on the West Somerset Railway with ex-GWR 0-6-0PT No. 6412 taking the leading role.

Left: The following year on 8 May 1978 No. 25 leads the side tank near the start of the climb to the exchange sidings. A couple of out of service engines can be observed beyond the gantry on the right. Pleasingly both engines have been saved for posterity, the pair now residing at the Scottish Industrial Railway Centre at Dunaskin, near Dalmellington in their native Ayrshire, although rather sadly No. 8 is in a rather woebegone state having attracted the attention of metal thieves in recent times.

Steam finally bowed out at Polkemmet in 1980, the NCB then hiring Class 08 diesel shunters from BR to handle the internal traffic. However the previous year management decided to honour steam's presence down the years by cosmetically restoring Andrew Barclay works No. 1175 in a light green livery with yellow lining, albeit with the wrong works plates affixed to the sides of the cab, those from Andrew Barclay works No. 1296. The name 'Dardanelles' was mounted on the running plate, a term by which the pit was often known locally, this naval battle raging at the time the shafts were being sunk in 1915. It was then mounted on a plinth outside the main entrance where it was recorded on 22 June 1982. The winding wheels above the upcast shaft also feature prominently, the concrete casing around the tower assisting the draught from below ground. Unfortunately the 1984 miners' strike sounded the death knell for the colliery, with no more coal being wound thereafter, and when most of the buildings were subsequently cleared in 1987 the locomotive was moved the short distance to Polkemmet Country Park, where it remains on display.

Kinneil Colliery at Bo'ness (Borrowstounness to give the town its rather cumbersome ancient title) by the southern shore of the Firth of Forth was developed by the NCB between 1951 and 1956. In 1960 272,000 tons of mainly high-grade coking coal was dug out from beneath the firth, a time when 1,057 workers underground, supported by a further 185 on the surface, depended on the mine for their weekly wages. From 1964 the coal from the Valleyfield take at Newmills, on the opposite side of the firth in Fife, was also hoisted up the shafts at Kinneil after the two had been connected underground. However in 1978 the direction of the conveyor belts were reversed and linked to the Longannet complex of mines in Fife, the coal then emerging by Longannet Power Station. This meant the end for the colliery sidings at Kinneil and as far as BR were concerned also the connecting branch that diverged from the Edinburgh to Glasgow Queen Street line at Bo'ness Junction, two miles west of Linlithgow. Passenger services along the branch had already been withdrawn in May 1956. As regards the administration of the colliery, because of its geographical isolation at the northwest corner of West Lothian, it never came under the same area management as the other Lothian collieries until July 1973, when the former Scottish North and South areas were combined to form the Scottish Area.

The engine depicted on these two pages, Andrew Barclay 0-4-0ST works No. 2292, NCB No. 21, built in 1951 with 16in. x 24in. cylinders and 3ft 7in. wheels, was the duty engine on 12 May 1972. The legend on the saddle tank shows the engine assigned to the Scottish North Area, evidence of a repaint since March 1967 when the area was formed by the amalgamation of the old Alloa and Fife areas with part of Central. On this page a monumental column of black smoke has just erupted from the chimney while on the right it runs sedately towards the colliery yard. The empty wagons on the left would slowly make their way to the far end of the colliery where a traverser, incorporating a weighing machine, directed them to the appropriate track for filling. Before coming to Kinneil in 1971, No. 21 led a somewhat peripatetic existence, spending its earlier years in Fife at Nellie (Lochgelly), then at Bowhill (Cardenden) and Rothes (Thornton) collieries before a move to Manor Powis Colliery, Causewayhead, Stirlingshire, in 1966.

No. 21 ventures out of bounds, pushing hard at the rear of a coal train led by BR Class 20 diesel-electric No. 20099, heading away from the colliery up the steep incline towards Bo'ness Junction on 24 May 1974. A few minutes later No. 21 returned light engine. Strictly such moves were not in accordance with the rule book. The Firth of Forth can be glimpsed above the train along with the scarred industrial landscape of Grangemouth.

Opposite: A smartly turned out NCB No. 41 (with no area designation), Andrew Barclay 0-4-0ST works No. 1107, bides time outside the two-road shed at Kinneil on 12 April 1971. A less powerful locomotive than No. 21, it had 14in. x 22in. cylinders and 3ft 5in. wheels and hence was not quite as popular, especially at times of inclement weather due to its open back cab. When new in 1907 it was supplied to the Fife-based Lochgelly Iron & Coal Co. Ltd's Minto Colliery as their No. 13. It was renumbered 41 by the NCB while still at Minto, before transferring its affections to Glencraig Colliery, Lochgelly, in 1956 and Kinneil in 1967. It was broken up in 1972.

Below: Driver James Spiers leans out from the cab of No. 21 while his mate (does any reader recognise him?) keeps a firm hold of the handrail. The cast oval Andrew Barclay identification plate stands proud.

Above: The end of the shift draws nigh on 23 May 1974 and James Spiers throws out the remains of the fire before bedding the engine in the shed for the night. Following retirement, No. 21 returned to Fife and is now in the care of the Kingdom of Fife Preservation Society at Kirkland Yard, Leven.